AF730860

The Spirit of Collaboration

Host Youth for Travel, Transition, Service, and Connection

Marian and James Kozlik

PLATYPUS
PUBLISHING

1st edition 2025

Cover photo from WWOOF: https://www.wwoof.net

Another short book from www.inspiredbookwriters.com

This book is dedicated to our good friend a naturopath, Zone Therapist, and fellow "barefoot gardener" - Marlenea La Shomb (April 11, 1958 — January 2, 2025)

Worldwide Opportunities on Organic Farms—USA... *WWOOF!*

Marlenea La Shomb
BAREFOOT GARDENING

Profile of a Permaculture Cooperative Right Here in Paradise Valley, MT

WWOOF is not a hound dog—it's a worldwide organization and effort to link visitors with organic farmers, to promote an educational exchange, and to build a global community conscious of ecological farming practices.

WWOOF was founded in 1971 in the UK, and is one of the world's first educational and cultural exchange programs. Today, WWOOF is in more than 132 countries around the world (and growing), with a wide range of farm-stay opportunities. WWOOF programs operate independently in each country, so please contact individual WWOOF organizations directly for the most accurate information on WWOOFing in the locale of your choice. This article highlights a local WWOOF farm in Emigrant, Montana.

WWOOF-USA is a proud member of the Federation of WWOOF Organizations, a democratic group with more than 40 national WWOOF members. For more information on the Federation, or on WWOOFing in other countries, please visit WWOOF.net.

Visitors, or WWOOFers, spend about half of each day helping out on a host farm, learning about organic farming and sustainable agriculture, and receiving room and board during their visit—with no money exchanged between hosts and WWOOFers.

The WWOOF-USA Host Farm Directory lists 2,386 organic farms (not necessarily USDA certified organic) and gardens across the country. Any farm, community or garden project using organic methods that would like to participate in a cultural and educational exchange may join the program. Anyone 18 or older, regardless of experience, may participate.

WWOOF farms offer a variety of educational opportunities, including growing vegetables, beekeeping, building straw bale houses, working with animals, making wine, and much more. With farms in all 50 states, the U.S. Virgin Islands and Puerto Rico, there is something for everyone.

THE SNOWBIRD AGRICULTURAL LAND TRUST IN EMIGRANT, MT

James and Marian Kozlik operate a WWOOF farm on Capricorn Drive in Emigrant, Montana, a 4-acre rural home site. They have developed an educational permaculture homestead for people interested in self-reliance and living in community, sustainably and off the grid. In exchange for free room and board, they offer plenty of chores to choose from and plenty of down time to explore the beautiful Yellowstone area.

Some of the Kozlik's upcoming projects include creating

Marian & James Kozlik

another hoop house, developing more bee and insect motels, gardening, and creating terrace garden beds. Field trips might include

a trip to a local monarch butterfly way station that releases monarchs. You may help retrofit existing buildings, expand and maintain current gardens, learn interpersonal skills, learn about land use, planning and design, create microclimates, forage for wild edibles, and maybe take part in community activism.

James recently had this to say, *"This is our fifth year with the WWOOF program. We get people from all over the world: Japan, Belgium, France, and other countries. It's quite a mix of people, age mostly early-to-late 20s, some early 30s. Our eldest request was from a 70 year old! Even though we are a small homestead, we have been recognized by the organization as a certified "farm." The Permaculture motto is care of earth, care of people and sharing surplus. We get to do that with this program. The payoff is meeting all the wonderful young people and world-wide travelers! This small service has enriched our lives."* ■

Contact James and Marian Kozlik at (406) 220-1563. Stays are limited to one or two weeks. Participants stay in a private room. Applications and registration for your stay is done through WWOOFUSA.com. Membership fee is $40 for a single and $65 for a couple and it is valid for one year.

Reviews

Jim and Marian offer a heartfelt, intimate picture of the joys and opportunities experienced through a permaculture garden setting. By sharing their permaculture knowledge with respect for all life they connect us with the bigger vision of possibility for one's self, humanity and the earth. Their journey is inspiring!

Mona Lewis, President Paradise Permaculture Institute

This book features the dedication of of a loving couple who have devoted so much to the World Organization of Organic Farming, opening their property, their time and most importantly their heart to volunteers who are interested in living in harmony with Mother Earth and practicing sustainable farming. It is delightful to read the vivid descriptions of living with and from the land, cultivating it, ripping its bounty and making a life of purpose out of it. These same vivid descriptions open a window to the heartfelt moments on the farm where hosts and Woofers learn the craft of enriching the land to harvest equally rich produce: the sights, sounds and scents of this scenery. Further moving into the aromas and flavors of the kitchen where the produce is stored and cooked as hosts and woofers bond over these activities and continue the learning process. I could not leave my chair turning one page after another, till I reached the back cover. With heartfelt sincerity,

- Teda Kokoneshi, LMHC

WWOOF?!? What in the world?!? When first reading about or coming across the Worldwide Opportunities on Organic Farms organization, a person can be amazed or intrigued or even motivated to become a participant. Sure, the organization itself publishes information, but how does one REALLY know how it works? Or if someone would like to possibly become a host, or even a WWOOFer, what is that experience REALLY like? Jim and Marian propose that they decided to write and offer this book as a resource, a challenge, a how-to if you will about what WWOOFing REALLY means to them. They absolutely deliver on that promise! This beautiful piece provides the reader with a true sense of what it means to share knowledge, share food, and share a home. You really get a sense of the original expectation versus the current goal and plan for being a WWOOF host. That journey was one of years for them, and they were able to condense that experience in a jam-packed work that will make the reader fly to the WEOOF website while pulling out a journal to start creating their own hosting plan. A heartfelt thank you to Marian and Jim for bringing us along in their adventure, and opening eyes and hearts to what WWOOFing is at its core. This particular series of anecdotes inspires a person to think about how they will deeply and profoundly connect with a person...beyond the foundation of how to farm organically. Reader beware: you'll likely be hosting a WWOOFer yourself after reading this book!!

Jennifer Carrick, B.A., M.Ed., CTRP, Micro-Homesteader in rural Central Ohio

In the closing words of his classic book, The One Straw Revolution, Masanobu Fukuoka states that "The ultimate goal of farming is not the growing of crops, but the cultivation and perfection of human beings."

So it is in permaculture as well. It is not just about the growing of food, the harvesting of rainwater or the construction of natural buildings. It is ultimately about our emotional and spiritual journey as human beings as well.

Somewhere in those great books out there, it says that we have been created in the image and likeness of 'The Creator' or that we have been endowed as human beings with the qualities of honor, reason, and responsibility. If this is so, then as creative and honorable people, why don't we just go ahead and create beauty and wonder around the world? We are going to do something with our days anyway. So, why not do something to create beauty and wonder instead of being inactive, consumptive and unconscious?

This is the choice that Jim and Marian have been making for decades. And who they have become and what they have created reflects a clear, solid, quiet and loving confidence. They have been walking the path toward greater emotional and spiritual maturity for decades while also working to leave their small piece of the planet in better condition than how they found it.

It is no wonder that Jim and Marian are a joy to be around. They are living an honest and authentic life with a warm willingness to share and educate while growing some of the healthiest food anywhere.

This book is a heartwarming read and loaded with practical insights for hosting a WWOOFing project. It is also a great look into how to live an authentic life. It is the real deal.

Many thanks Jim and Marian for your life's work.

- Bill Wilson,Cofounder of Midwest Permaculture

Foreword

WWOOF is more than just a way to learn about organic farming and sustainability—it is a profound experience of connection, learning, and shared purpose. Over the decades, WWOOF has grown into a global movement, fostering relationships between hosts and volunteers that transcend borders, backgrounds, and generations. This book, A Spirit of Collaboration, beautifully captures the essence of that experience through the heartfelt stories and reflections of hosts who have opened their homes and hearts to WWOOFers from around the world.

As the Executive Director of WWOOF-USA, I have had the privilege of witnessing the countless ways WWOOF transforms lives. Volunteers arrive seeking knowledge about organic farming, sustainability, and permaculture, yet they often leave with something even greater—lifelong friendships, a deeper connection to the earth, and a renewed sense of purpose. Hosts, in turn, are enriched by the energy, perspectives, and enthusiasm that WWOOFers bring to their farms and homesteads. It is a relationship of mutual learning, growth, and gratitude.

In these pages, Marian and James Kozlik share not only their experiences as WWOOF hosts but also the philosophy and spirit that make WWOOF so unique. Their stories are a testament to the power of community and the idea that we are all stewards of this planet, responsible for nurturing the land and one another. They remind us that hosting is about more than teaching sustainable farming or homesteading techniques—it is about fostering a culture of kindness, respect, and shared responsibility.

As you read this book, I hope you will be inspired by the journeys of the WWOOFers and hosts whose paths have crossed in meaningful ways. Whether you are a seasoned host, a prospective WWOOFer, or someone curious about the WWOOF movement, may these stories encourage you to embark on your own journey of connection, discovery, and collaboration.

With gratitude,

Sarah Potenza Executive Director, WWOOF-USA

Introduction

A Spirit of Collaboration is our story, a heartfelt guide for anyone looking to reconnect with nature, build meaningful relationships, and contribute to a more sustainable way of living. After over a decade of hosting WWOOF volunteers on our homestead, we've learned so much about the power of collaboration and the joy of sharing our lives with others.

Through this book, we share our personal experiences—moments of connection, challenges we faced, and the growth we've experienced. Working with the land has taught us the beauty of growing food, the art of cooking from scratch, and the importance of living in harmony with the earth. Hosting has shown us how to nurture true connections—not only with our guests but also with ourselves and the world around us.

We hope this book can serve as a guide for others, offering inspiration and practical advice for those who want to create their own spaces of learning, collaboration, and renewal. For us, this journey has been about more than farming; it's about fostering community and helping shape a cultural shift toward living with purpose, gratitude, and respect for the earth. We've given what we could, and in return, we've been gifted with the energy and exuberance of those we've welcomed into our lives. This is what A Spirit of Collaboration is all about—coming together to create something beautiful and meaningful.

Chapter 1

Yoga, Knapweed, and Sunrise

Starting the day and developing a passion

> What is a weed? A plant whose virtues have never been discovered.
>
> Ralph Waldo Emerson

We love sharing our homestead with wwoofers. It is nestled on a bench overlooking the Yellowstone River. After leaving the hustle and bustle of Chicago, we embraced a simpler, self-sufficient life on four acres of Montana's wide-open spaces.

Typically, we like to start our day off with a little yoga. Over the years we did a lot of different practices. And there have been many of our volunteers (wwoofers) who also like to do yoga. We'd take turns leading a session in our living room. Some chose to do it alone outside on a deck, and we sometimes prefer to do ours alone too. There had been times when wwoofers attended Kundalini yoga classes in town with us. This gave them opportunities to experience our local community.

Sometimes the yoga transitions into Sava (service) the hard but satisfying work of pulling the invasive knapweed. Even though it is a bee attractor, it will take over a field. Since we live in ranch country we want to contribute to the well being of our neighbors. We do not spray, instead we physically pull it out of the ground.

We scour the land with gloved hands, shovels and a keen eye looking for the weed. It's not a glorious type of contribution yet we observe that most wwoofers are willing to take part in this activity of service. This is one of many gifts we receive from our wwoofers. We like to point out the flora of the native plants that grow low to the ground and are not noticed from a distance.

In time, we've noticed the help of a natural ally—the weevil, which burrows into the knapweed roots. Even better, Sheep and goats will eat knapweed. For us an "ocean" of Blue Flax has taken over large areas, bringing a breathtaking sea of blue to our land each spring and helping to hold back the knapweed.

Other tasks may include watering, slug patrol, weeding in the garden, picking raspberries and goose berries, picking greens for the green drink, or continuing a project that was started the previous day.

Our days are guided by the rhythms of the land and community. When the Arcosanti bell rings it's time for a check-in. The early morning tasks allows us to start each day connecting with nature.

Youtube Link — "Yoga For Gardeners"

Our incentive for having yoga and early morning chores:

- To set a routine for the day's garden work.
- To develop a ritual of connecting with nature and the garden.
- To welcome in the sun and the new day.

Implementation Journaling Questions:

1. Create a plan for how you might want to start the day with your WWOOFers
2. Create an early morning routine and/or ritual that you would like to offer to your WWOOFers to partake in at the start of the day

Chapter 2

Check in and Green Drinks

Feelings, emotion, compassion and empathy

One of the most meaningful rituals we've created on our homestead is Check-In. Every morning, we gather in the kitchen where Marian prepares her signature green drinks. These are a blend of greens and flowers from the garden, fruit like bananas or apples, Ayurveda powders, and hemp seeds. Most of the time, they're delicious; other times, they're simply packed with health!

Picking greens becomes a lesson in plant identification, as well as, learning the nutritional value of the plant. We take our wwoofers around the garden so that one of their morning tasks is harvesting wild and domestic greens for our morning green drink. Not everyone partakes in drinking these concoctions, that's okay, too.

With our drinks in hand, we move to the sunroom, soaking in the garden and mountain views, often accompanied by the sound of chimes.

We begin with a Serene Mind meditation, a practice that has become central to our mornings. One of us guides us through it—starting with the breath, a brief reflection on a thought or emotion, and an observation of whether our mind is caught in the past, future, or present. Together, we focus on a flame traveling from the third eye to the center of the brain.

This is a quiet and centering moment. Then sipping our green drinks and gazing out at nature, we connect deeply with ourselves and each other.

The heart of our mornings, however, is "Check-in." For us, it's a powerful practice to shift from a suffering state to a non-suffering state. WWOOFing is fundamentally about relationships—whether it's with ourselves, others, or the earth. During "Check-in," we share openly our emotions. We go inward and observe whether joy, sadness, confusion, or possibly something deeper may be surfacing. Over time, as trust grows, these moments become richer, more vulnerable, and more authentic.

Living on this land has taught us so much about slowing down, connecting, and creating. It's here that we've come to understand how our inner state influences the energy of everything around us. The farm offers time to contemplate, to witness how the simple act of creating in the garden mirrors the creation of our inner world. Lately, we've embraced the liberating awareness of "I created this," whether in the garden or in life, and it's a deeply fulfilling experience.

We believe all of us have a choice—to live in the "old world" of conflict, competition, and disruption or a "new world" of cooperation, compassion, and collaboration. As we free ourselves from what holds us back, we cultivate a greater capacity for awareness, observation, and simply being.

Our journey has been shaped by decades of inner work and wisdom from many teachers, including Marshall Rosenberg, Byron Katie, and our Mutki Gurus Sri Krishnaji and Sri Preethaji. Since the late 1960s, we've both been seekers, always striving for a deeper inner connection with ourselves, one another and the world. This chapter has been a reflection of how those teachings have come to life in our daily rituals, bringing purpose, relevance, and joy to our days on the homestead.

There have been times when a wwoofer has come carrying a huge burden. We felt it. When you can feel the other is when true compassion is present within us. That's when a heart felt connection occurs. We've witnessed how time in nature has been a great antidote for someone in a suffering state.

Our incentive for creating morning Check-Ins:

- To develop closer relationships while working and growing together.
- To develop fluency in expressing feelings and needs.
- To support one another in building a community environment.
- To meet the need for open and honest connection.
- To inspire a passion to connecting with Mother Earth.

Implementation Journaling Questions:

- Create a plan for how you might enhance your relationship with your WWOOFers

- Create and prioritize the tasks you would like to have completed for the day

Chapter 3

"Getting to Know You" (Breakfast)

Connection and opening our hearts

> "Love is where you feel the other to be a gift from the universe. You live in the knowing that what they bring to your life is irreplaceable."
>
> Sri Preethaji

Breakfast isn't just a meal—it's the start of something deeper. By the time we sit down together, the connection has already begun to take shape, starting with check-ins and slowly building over time. It doesn't happen all at once. Instead, it unfolds naturally—whether it's in the garden, during meal prep, over shared meals, or on walks and hikes. These small moments become the spaces where we share stories and perspectives.

Sometimes, people show up carrying their own burdens, and it takes a while for those things to come to the surface. Other times, we've realized we weren't fully ready to welcome someone because of the state we were in ourselves. But every visit has its own rhythm, and connections tend to develop in ways we never expect.

In the Hut, a small single room, there's a little library with books on gardening, homesteading, novels, and Permaculture resources. We encourage wwoofers to browse through the materials, and asking if they've looked at any of it often sparks meaningful conversations. It's a natural way to explore their reasons for being here, and it helps us share what our area has to offer, especially for outdoor recreation. Morning check-ins are also key to understanding their goals and forming a stronger connection.

Being in the garden together is one of the best ways to build those relationships. A good way to begin knowing one another is by giving a tour of our growing areas. We do plant identification and acknowledge the medicinal or nutritional benefits of each plant. By having them harvest these greens, for our morning green drinks, it draws their attention more fully to each plant and connects them to the garden.

Early on, we had to shift our mindset. It's easy to think, "They're working for us, so we can focus on something else," but we've learned that wwoofers often need clarity or guidance. Sometimes, we like to challenge them—show them the plants, tools, and materials then let them create a project.

View of garden section with bee and butterfly attractors and wild edible and medicinal plants. The Russian Olive tree considered a weed and illegal to sell is for the protection of birds. They flock to it!

Whether it's a garden plot, worm farm, or irrigation, creating a part of the homestead has been empowering for everyone involved.

Our hosting for this last decade has had a great impact on our lives. Not only have our experiences enhanced our gardening enthusiasm, we also discovered simpler techniques to be better hosts. We have a greater appreciation for our youth and their desire to serve and inspire change in our world. Perhaps the greatest impact has been in each of us growing personally, along with all of our relationships.

We are more clear with setting boundaries, more open and honest with self and others. We have learned to appreciate honoring our limits and accept our shortcomings. Being hosts has allowed us the opportunity to connect without judgements or projections. We can witness these suffering states and move on to connecting with unconditional love. Loving what is, like when the red poppy got chopped down. The two lady wwoofers were too much fun to let a plant get in the way of our relationship.

Meal prep and meal time is a great time to get to know our wwoofers. There is something about everyone taking part with the food. One of our first wwoofer's announced his expectation that we cook and he works. Soon his mate would help us prep lunch. She would get salads together, we would be getting the main portions prepared. It wasn't long after that he started to join us. Every main meal became sort of a celebration. Soon he was making the best European chili "this side of Texas".

We had a wwoofer who spent just as much time in the kitchen as she did in the garden. We sure did not mind cleaning up after her!

There was the chef that came from a ski season at Big Sky. He developed a non dairy Creme brûlée. It took a few tries but we didn't mind all of the delicious taste tests.

The Japanese sushi chefs came from BC, Canada after a ski season there. They said we ate a lot like them, they prepared three days of vegan sushi!

Another time we had a traditional African meal. The times we spend in food prep and eating has been precious. A time spent as community in the "community kitchen".

Our incentive to encourage considerations for a hearty and varied breakfast is:

- To start the day with good nourishment.
- To prevent a mid-morning blood sugar attack that drains a WWOOFer's energy
- To encourage WWOOFer participation in preparing the breakfast

– whether it's setting the table, grating hash browns, making toast or picking greens for the green drink

- To have a prayer of gratitude before eating. A great way to start the daily meals is to take turns leading the gratitude prayer

Implementation Journaling Questions:

1. Create a plan for how you might create an atmosphere for a "community kitchen" with your WWOOFers
2. Consider how to inspire community food prep so it carries over to each meal...otherwise the host will always be in the kitchen

Chapter 4

Relationships

Personal growth and wisdom

"There is no greater agony then bearing an untold story inside you."

- Maya Angelou

There is always weeds in the garden. Some we leave, some we pull and some become mulch and compost to recycle. Weeding can be a turnoff for woofers. "I thought I was going to learn gardening skills." (Sic)

We have several observations, some may be true and some might be perceived or projected by us. So we have to weed those thoughts out of our mind and connect in a meaningful way with our farm guest WWOOFer.

Hosting wwoofers has been one of the most meaningful parts of our lives here on the homestead.

Jim - I'll never forget a particular wwoofer visit—her calm, present energy left a mark on me. Marian was away in Florida, and one day I found the wwoofer working quietly near the chimes, her focus entirely on the earth. I was inside, lost in real estate work, when I realized how absent I'd been. Marian wasn't there to connect with her, and neither was I. Something pulled me outside, and as I stood there watching her, I felt shy, unsure of how to approach.

I finally walked over and asked, "How are you doing back here? Are you feeling deserted?" Her response stayed with me: "Are you kidding? The chimes, the view, working with the earth—it's a blessing. I'm just being! Thank you for asking. I'm doing good." That moment reminded me how rare it is to truly just be.

In the early days of hosting, I was a bit of a chatterbox, eager to share our story—how Marian and I started homesteading and everything we'd learned. But over time, I realized connection wasn't about sharing our story; it was about listening to theirs.

One day we (the aformentioned wwoofer and I) hiked Emigrant Gulch together, scrambling through ravines and boulder fields until we reached a peaceful spot to rest. We sat separately, soaking in the vast stillness of the mountains. It wasn't until the hike back that our wwoofer opened up about her life—her move from Pakistan to the U.S., her training in acupuncture and Chinese medicine. Later, she gave me a suction cup treatment for my sore back (which worked wonders) and sorted a dusty collection of Chinese herbs we'd inherited, labeling each one and writing out its uses. It was a gift we didn't even know we needed.

Marian - Let me share one experience that speaks to our opening statement. We had just returned from four days in the CCU (critical care unit) of the Bozeman Hospital, the previous day. Jim had a near death experience from a sepsis infection. The doctors could not keep his temperature down so they induced a coma, paralyzed him and placed his body in an ice bag. Jim survived. This particular day Jim was out with his brother and sister-in-law.

I was home alone and heard a knock at our front door. As I arrived at the door a beautiful young woman stood smiling at me. I was somewhat disoriented from days of sleepless nights. WHO COULD THIS BE? Then it dawned on me that we were expecting a wwoofer, a guest to help in the garden. The door was open only the screen came between us. We looked into one another's eyes and neither of us said a word. It was as if our hearts did all the talking. I opened the door for her and she gracefully entered. We embraced as if we were the closest of kin. Then holding each other at arms length our eyes connected and our hearts sang. There was no awkwardness only communion.

Jim - "One day, while I was digging in the soil with bare hands and feeling the earth and thanking the worms, something occurred to me. While our "wwoofer" (an acronym for our volunteers) was building a raised bed for next years potato patch, I realized that I was spending most of my time on the phone selling real estate. I looked over at Marian, planting. Then I looked at our wwoofer, my emotions lead to some tears, because I knew that a heart to heart connection was missing. We were a short distance from each other. I knew I was hanging on to perceptions and judgements about someone I really didn't know. I was thinking that he took advantage of our "flexible schedule". His remote work was the priority over time spent helping out. My business calls took priority over forming a meaningful relationship. I made a statement, "While I have been digging, connecting with earth, I realized that I haven't taken the time to connect with you!" What happened after that was a liberation, from my perceptions of him. And that is when our relationship blossomed."

Every wwoofer who's come through has brought us gifts, whether tangible or in the form of lessons we've learned together. There was a young woman from Utah, with a homesteading background. She taught us new skills, like building worm towers for our garden. She was a whirlwind of energy, cooking, baking, and tackling projects with an intensity we could barely keep up with.

Then there was a recent grad, an engineer from New Mexico. He stayed far longer than planned, initially testing our patience with his meticulous approach to projects. He designed and built a water catchment system for our gutters and a beautiful pathway in our terrace garden. By the time he left, we'd grown close and had gained a deep appreciation for his precision and thoughtfulness.

Matt's rain barrel project waters an Apple Tree, Plum Tree, Asparagus patch, Sweet Marjoram, Lavender, Poppies, Horseradish, and much more.

Not every moment has been perfect, but even the mishaps have been meaningful. For instance a young woman still going to college, mistook our beloved red poppy plant for a thistle and cut it down. The check-in the next morning was intense as we worked through forgiveness, guilt, and disappointment. We assured her the plant would recover, and it did—coming back stronger and more beautiful the following season. We sent her a photo, and she reminded us, with a touch of humor, that she knew what she was doing after all.

Her friend and roommate, helped us save half our apple crop during a fierce windstorm. Together, we built a windbreak, and when they left, their car was packed with plants and smiles. Later, we received photos of their dorm balcony garden—proof that their time here had taken root in more ways than one.

These connections are the heart of what we do. Every guest has shaped us in some way, teaching us about collaboration, resilience, and the beauty of shared moments. The relationships we've built over the years are what truly define the spirit of our homestead.

We are continually amazed by the many precious contributions our wwoofers have made. They've built bee condos, worm boxes, an outdoor shower area, a solar array, garden beds, and even painted a mural on the front gable of the house. There have been endless repairs and creative projects that bring our homestead to life, and each visitor leaves their mark.

Our intention is to empower our WWOOFers and connect with them. We observe our state so we might connect with equanimity.

- To enjoy the flow of the day doing projects and finding solutions

- To learn more about their life and them on a deeper level
- To let them co-create on the homestead and connect with the Earth
- To utilize the materials that are on-hand
- To inspire a permaculture ethic. A sense of joy vs. work

Implementation Journaling Questions:

1. Make a plan for how you might create an atmosphere of creativity with your WWOOFers
2. Consider what projects might enhance the homestead. Discuss this at check in and see what the WWOOFers might like to volunteer to do. A project to benefit the overall garden design.

The raised beds in our hoop house were built by a WWOOFer and it has produced abundant tomatoes, cucumbers, peppers, eggplant, watermelon, basil, Swiss chard, collards, and ashwagandha. The middle raised bed started with straw bale and we grew tomatoes in them the first few years. Eventually we filled it in with dirt.

Chapter 5

Let's Plant a Garden

Domestic plants and weeds in collaboration with the whole

If we look at the chart from the book Together Resilient by Ma'ikwe Ludwig (Figure 1) we can use the columns referencing the extreme competitive culture, the sustainable cooperative culture , and the extreme cooperative culture and apply many of the references to gardening. The competition of weeds - working in cooperation with the weeds (and what they provide), or only collaborating with domestic plants.

Comparing Cultural Worldviews
and the behaviors that come from them

Extreme Competitive Culture	Sustainable Cooperative Culture	Extreme Cooperative Culture
Compete with others	Cooperate, including collaboration with allies	Cooperate within; collaborate only with others your group fully agrees with
Seek advantage and winning	Seek understanding and effective action	Seek attention as connection
Have skill? Use to dominate	Have skill? Teach with discernment	Have skill? Give away indiscriminately
Loudest voices win	Collaboration/consensus with discernment	Consensus with no discernment

Extreme Competitive Culture	Sustainable Cooperative Culture	Extreme Cooperative Culture
I-oriented (individualistic, focus on self)	We-oriented (communal, focus on self in balance with others)	Us-oriented (hyper-communal, self subsumed to group)
Independence encouraged/celebrated	Interdependence encouraged/celebrated	Codependence encouraged/celebrated
Dis-integrated	Integrated with differences valued	Individual needs/ strengths lost
Capitalize on circumstances	Empathize with circumstances	Pity circumstances
Protect (resources and emotions, with no risk)	Share (resources and emotions, with boundaries)	Share (resources and emotions, without boundaries)
Make others responsible	Recognize personal and collective responsibility	Over-own personal responsibility
Differences threaten me	Differences are interesting	Differences threaten the group
Narcissism based on being "the best" and not needing to care	Not narcissistic: self is valued member of valued team	Narcissism based on emotional neediness met by group
Systems serve me	Service to others	Martyrdom

Figure 1: Chart from the book Together Resilient by Ma'ikwe Ludwig - Comparing Cultural Worldviews and the behaviors that come from them.

Our garden is created one plot at a time. Each plot is different and when we observe the plots from season to season there came a time when we learned to appreciate each plot for what it contributed to the whole of the garden. We eventually transitioned from feeling overwhelmed and obsessed with the weeds. We also transitioned from dictating where each plant would grow. We settled into a more sustainable attitude of cooperation with the weeds, the plots, and where some plants decided to move too. We became fascinated to see how the Aspens moved into our raspberry patch and watched as the raspberries scattered into other areas. This is the same with much of our garden, for instance look at the picture of the Blue Flax in chapter two. Or the photo of the bee and butterfly attractors with wild edibles (chapter three). That used to be manicured every year for domestic plants.

We had a fetish concerning quack grass. A very hardy grass around here with a root system that never goes away. During our "springtime overwhelm phase" of our gardening journey, we were digging out the grass to make a plot for some vegetables. As we were digging we started noticing how many more worms the plot had than plots that were made barren in the fall. The earth was still moist and rich. We considered how the grass kept the soil from being swept away by the wind.

Marian started laying the grass out to dry and eventually utilized it for composting or mulching. From then on we had "the spirit of collaboration" with our garden. That photo in chapter three, as I look at it I can see the plot where we made our transition (sounds heavenly). When that photo was taken Jerusalem Artichokes took it over. Through the years it has been a plot for many cultures to enjoy, lettuce, tulips, mint, parsnips, onions, beets and more. The grass still comes up. This year a wwoofer layered it with chopped leaves, compost, composted soil and straw. The grass will still pop through, the worms will have more to chew, and whatever grows there will flourish!

Our mountain valley season starts in May. Prior to that we will have started seeds and nurtured them in the garage winter greenhouse, sunroom and bedroom windows. We use to have lights and started seedlings as early as March. The garden work is different depending on the month a WWOOFer comes to help.

Our Objective is:

- To know our zone and when to start seeds
- To dead head what was left from the previous season
- To learn the best transplanting days
- To know about WWOOFers garden experience

Implementation Journaling Questions:

1. Create a sequential to-do list for your WWOOFers
2. Consider your zone, when to and where to plant. You might make a map with your WWOOFers indicating where you want the crops to be planted. Create a layout of your garden.

Chapter 6

What are our WWOOFers doing?

Projects, tasks, field trips and accomplishments

This chapter captures the rhythm of our days with WWOOFers and the ever-changing nature of farm work. We often start with a plan, but things shift —especially when we, as hosts, don't always agree on what should take priority. Over time, we've learned to keep a list of ongoing projects that always need attention, the daily homestead tasks.

Slug patrol is one of those tasks, and it never fails to amuse us. One WWOOFer (we won't name names!) took great pride in his method, sitting on a stump, tipping over another, and filling an ash can with slugs. "I got twenty today!" he'd announce. The irony? He was picking them from a garden bed with nothing growing in it. The cabbage patch would have been a better choice, but it required a more careful approach.

Watering is another essential morning task, and we prefer to do it by hand. It's not just about giving plants water—it's about walking, talking with and hugging the garden; noticing what's thriving, what's struggling, spotting pests, pulling weeds, or picking something ripe. Laying the hose down to soak an area while we make observations. Done right, it takes a couple of hours. So when a WWOOFer proudly says, "I watered everything!" in twenty minutes, we know it's time for a lesson.

We once had an irrigation system. Our oldest son visited, watched us struggle, and ran off to Home Depot to get the supplies. A WWOOFer later took up the project and finished the setup. It worked well for a while—until we realized something was missing. The ease of flipping a switch took us out of the garden. Instead of connecting with the plants, we got distracted doing anything but focusing on the garden first. We lost something in the process.

So we went back to the old way, adding water gauges to help us determine which beds actually need watering. Now, it's a morning task for our WWOOFers to check those gauges and decide what needs attention. It keeps us all in tune with the land, and that's what really matters.

We keep a running checklist of garden tasks and bigger projects that need doing around the homestead. Some are routine—like splitting firewood, fencing, deadheading flowers, or harvesting—while others are bigger, long-term projects: hoop house repairs, water catchment, new garden beds, cooking cultural meals, worm farming, painting, and more. We generally let the wwoofers pick projects or tasks that they want to do. Sometimes they ask to create a project that will contribute to the homestead.

The outdoor shower project started with a simple request. A WWOOFer couple, staying in the hut, wasn't thrilled about walking past the sunroom in just a towel to reach our usual solar shower in the hoop house. They asked if they could build a shower at the far end of the property, and we happily agreed.

To make it feel complete, they wanted a proper stone pathway. So, we all piled into the old pickup and headed to a place to collect flat rock—right into grizzly country. Sure enough, as soon as we stopped, a grizzly mom and her cub strolled across the far end of the field. The WWOOFers, wide-eyed, stayed near the truck while gathering rocks, and keeping a nervous eye on watch.

A short hike turned into a cautious dance—stop, look, listen, then go. We shared bear stories, which they didn't find quite as funny as we did, but by the end, we had enough rock to finish the path. The shower is still a work in progress, but the adventure of building it is a story we won't forget.

Some projects on our homestead come from necessity, while others come from the creativity and skills of our WWOOFers. One summer, during an especially hot stretch, we had a couple staying in their small RV. The house stays pretty cool, so they often came inside for relief from the heat. While spending time indoors, they noticed that the walls could use a fresh coat of paint. Turns out, they had run a painting business while in college. They volunteered to take on a full interior paint job as their project—not only did it fulfill a real need, but it also kept them out of their sweltering RV. We got a professional-level paint job, and they got a more comfortable place to work. A win-win.

Then there are WWOOFers who bring their own project ideas to life. One guy, in particular, decided to build bee condos—small wooden shelters designed for solitary bees. He mounted them to the top of the garden fence, and we left them to see what would happen. The following year, we checked and found that two of the condos had been claimed by bees! Excited, we called him up to let him know that his project was a success. We joked about charging rent, and without missing a beat, he said, "Of course! The rent is honey."

We can go on with more stories. You get the idea, we love to see this transitional homestead be an experience of a creative process. Wwoofers leave their mark, so to speak. And if a project isn't completed there is always someone to carry it on.

Extra Curricular Activities

Now we want to share the value of field trips. We had been fortunate in the past to have some **field trips**. We have a neighbor in another subdivision that has put together a **"way station" for monarch butterflies**. We would call to find out if any hatches were happening. We'd take wwoofers over and visit the lab and even take a chrysalis to place on a milkweed. It would take awhile yet what an awesome experience to watch it come out of it's cocoon and start feeding on it's first "meal" and then fly off. We have gathered seed from wild milkweed and planted them in our garden.

Another field trip was done when two wwoofers volunteered to assist a woman at **Montana State University** planting specific field plants in a university **agricultural department** field. They had a very enjoyable day and the professor called us and was so grateful to have the day with them. Not only was this a learning experience, it also contributed to community.

We have made several visits to a local **aquaculture farm**. It was recently wiped out by a hundred year flood of the Yellowstone River. A **worm farm** was another field trip for wwoofers who got to experience the operation.

Some wwoofers partook in **community work** helping a locale lady plant her flower garden. Another assisted Jim doing a foundation for a **Habitat for Humanity** home in our neighborhood. Many wwoofers helped release Monarch Butterflies at the **Monarch Butterfly way station**.

There have been wwoofers who where short on cash and they got to work with a landscape friend of ours (who passed this last year). He **employed** them with landscape jobs. It got to be a joke whenever he would show up we'd lose a wwoofer for a day or two.

And we have had a few wwoofers **stay in the area** getting jobs, landscaping, a chef for a locale dude ranch, house keeping at Chico Hot Springs Resort, and with a concession employer at Yellowstone National Park.

Our intention is to empower our WWOOFers as if this garden and homestead is their community. For them to conceive the potential that they may want to be part of or contribute to a community. It doesn't matter if it's in an urban setting or a rural setting. What matters is the impact of being here and experience life differently.

- To enjoy the flow of the day doing projects and finding solutions.
- To let them co create on the homestead and connect with the earth.
- To utilize the materials that are on hand.
- To inspire a permaculture ethic. A sense of joy vs work.

Implementation Journaling Questions

1. Make a plan for how you might create an atmosphere of creativity with your WWOOFers. Discuss what the WWOOFers might like to volunteer to do and how the project will benefit the over all concept of the Farm/Garden.

2. Consider what materials they might need for their contribution. What tools they might need. Have they ever used power tools? What might you need to purchase to complete the project?

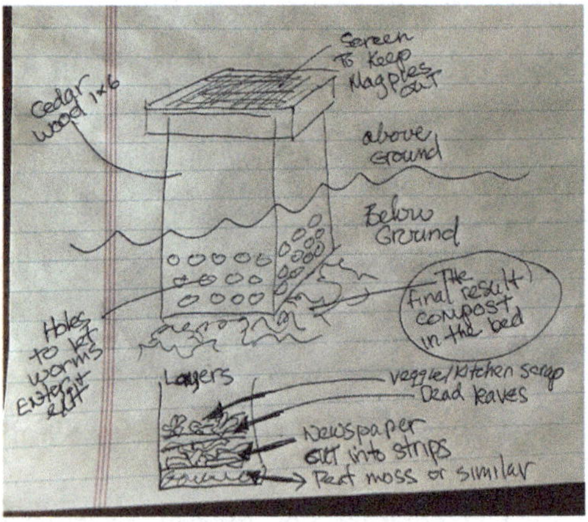

WWOOFer - Sydnee's worm boxes. We did a field trip to a local worm farm in Livingston where a couple had a huge operation in their garage and sold worm casings and worm tea.

Chapter 7

What's Cooking?

The cultural dinners and woofer specialties

We touched on the subject of cooking in chapter three. Typically as the day wears on there seems to be a natural gravitation towards the kitchen. Sometimes it is subtle a wwoofer might be reading a book or just hanging out in the sunroom. Everyone knows when dinner is to be served. In the past it seemed as if we were responsible for preparing a meal. This is no longer the case. In fact, our very first wwoofers offered to cook. It's become an organic process for wwoofers to participate in the meal prep. It establishes a sense of community. A real family like connection.

We have come to realize how cooking together is a way to demonstrate family traditions, culture and love. It's become an integral part of our hosting wwoofers. We talk, get to know each other better and morning check in becomes more personal. The host is expected to supply the food. As we get to know our wwoofers, we like to ask if they enjoy cooking. We inquire what's their favorite cuisine. If we get a positive response then we like to encourage them to "take initiative in the kitchen". We encourage them to be empowered, connect with the food, be grateful for what the earth supplies. A wwoofer might need ingredients for a special meal that they prepare and cook. We are willing to get what they need.

Cultural dinners can really facilitate inspiring conversations. For instance, we had a woman who did her Masters Degree from Indiana University on

Raptors. She went to a place in South Africa where there were Raptors. She was able to get her Masters program paid for by the University. She had to manage male Africans as part of the project and lived in the village. She made a traditional African meal that we ate with our fingers. It was simple and delicious. We had fun eating in the traditional way as the African villagers and we learned about Raptors. She shared photos of her crew, the village, and the mountain where the Raptors were located. It was such an experiential meal and discussion. We really got to know her, and she returned to visit us the following fall.

She and her partner stayed with us for a weekend. We were privy to her new life in Montana. They were living off grid, the pioneer lifestyle, which is how we lived at 8,000 feet in the Colorado Rockies. Their lifestyle matched ours and we traded goods. We had a wood burning water heater that we never hooked up. They had a hot tub and needed someway to heat the water. We talked about the wood burning stove we used to heat our earth sunken hot tub in Colorado. They shared stories of the Montana wild edible workshops they took. They had special high altitude corn seeds that they grew, and packaged to sell. In the end we traded a wood burning hot water heater in exchange for Montana wild edible books, corn seeds, beeswax, candles and a bison skull. More than anything we received an unforgettable connection of friendship. We made a dinner salad from the wild edibles in our garden, and a meal with wild mushrooms that they found. They brought bison steak that came from the annual Bison Hunt that takes place outside of Yellowstone N.P.. It is a controversial event that happens as a quirky way to manage Bison.

We won't get into it here, but that reminds us of a Wolf Foundation activist and friend of the Bison, who has spent time with us and even stayed here while she wrote one of her books about wolves. She sometimes joined us in the kitchen with us prepping dinners and making us specialities with her famous New Mexico chili peppers. We had many after dinner discussions that lead to continued visits when she was in the area. In fact we hosted a "dynamic" radio show with her (Meditation to Self Realization) for KGVM a locale public radio station. She stayed for a time taking care of our homestead, cat and garden while she wrote a book. And she developed a lecture that she gave in Gardiner, concerning Bison in Yellowstone N.P..

What's cooking is more about a recipe for developing deeper relationships. As Wwoof hosts we understand that nature plays a huge role in self development. The nature of food brings us together as gatherers, processing wild and domestic plants. It sustains relationships between the environment, people and what we eat. For some of our wwoofers it opens the door to connecting with elders who are not their parents. They can feel liberated to discuss who they are and what they believe in without criticism from their family "tribe" members.

One of our wwoofers from India, prepared Indian dishes. This met our needs because we were getting ready for a six month spiritual and personal development sojourn to Ekam, in India. She worked remote and some days she would be at the kitchen table finishing up her work. Much like our daughter does when she visits. What happened was in depth conversation concerning her work and marital situation. It gave us an opportunity to look at our marriage and appreciate the way we created a sustainable relationship despite our differences.

We had a small fourteen foot teepee who's covering finally deteriorated. It had bisons painted around the circumference. When we striped the covering off of the poles, we left the poles standing in the cone shape. We read a book that mentioned using copper, cones, special shapes and minerals that attracted energy from the universe to grow better gardens. It was pretty cosmic, "Secrets of the Soil" by Peter Tompkins and Christopher Bird (Books section #11). It's about "new age solutions for restoring our planet". If I remember correctly, anything growing north of a cone shape gains the benefits of the energy from a cone. Much like the mystery of the pyramid shape. So we planted the "Three Sisters" under the teepee poles. The special high altitude corn from our trade, previously mentioned, beans and squash. The staple of Native Americans, and boy did it produce!

What's Cooking is about being aware as a host. An awareness concerning our needs and the needs of the wwoofers. Food for thought is observation of our surroundings with people and nature. It's witnessing the time spent and being conscious of patterns that we have created and how they impact the other. Just like how the Aspens impacted the raspberries and how in turn the raspberries impacted the garden beds they moved into. Similar to the way our garden is impacted by the wwoofers that come and go. That is what's really cooking when we host.

Our intention is not to be the chef at a restaurant. If the WWOOFers are a member of this transient community we can inspire them to conceive of the potential that they may want to be part of, or contribute to a community. It doesn't matter if it's in an urban setting or a rural setting. What matters is the impact of being here and experiencing life differently.

- To enjoy the flow of the day which includes food preparation.
- To ask them for help in the kitchen - prep work, cooking, setting the table, gathering produce, even help with clean up.
- To inspire a joyful kitchen and enhance our connections.

Implementation Journaling Questions

1. Make a plan for how you might create an atmosphere of more personal connection with your WWOOFers. Food is a great bonding element.
2. A WWOOFer may wish to prepare their favorite meal. What ingredients could they need? Are you willing to supply everything they need?

Chapter 8

Flexibility on the Homestead

Enjoy your vacation – learn experientially – with mindfulness

> We have the choice to live lives of unawareness or make a conscious effort to impact the lives of others from deep thankfulness for what we receive.
>
> - Sri Preethaji

We happen to live in an area that has become very popular over the past thirty seven years. The past ten years has seen stupendous growth with new houses many with an income source for a vacation rental. People come here to see Yellowstone National Park. Or for fishing the Yellowstone River. The surrounding mountains with wilderness and national forest access make for great outdoor recreation.

Not to fool ourselves we realize that many of the wwoofers want to enjoy some vacation time. And that is why we have a flexible schedule. We require six hours work for two meals breakfast and dinner. Breakfast is usually around nine thirty in the morning and dinner around two thirty. We've had some people insist that they don't eat breakfast they would rather have lunch and dinner. We do not make that an option. They have a set up in the hut that they are staying in, for cooking later in the evening if they want.

One nice thing about dinner at two thirty is that there is still plenty of time to pursue a local hike, or jaunt into town. Some people want to camp for a day or two, we let them adjust their schedule to work more hours so they can take an additional day or two off. We generally take Sundays off.

We like our privacy. In the beginning we'd let them use the kitchen if they were cooking a late snack or dinner. Not anymore. A "community kitchen" concept is tough to do when the set up is a private residence. Especially when we have visitors.

What we have found is that hosting an educational type program has a lot of demands on the host. There's quite a balance of developing relationships, being clear with instructions for a task, and allowing a wwoofer to feel like family. In the previous chapter we discussed how being in the kitchen helps develop relationships. However, for relationships to have trust, there is a lot of inner work that needs to be acknowledged. When we (hosts and wwoofers) have private time to just be, that is when we have time to be introspective. There's time to contemplate how a task has been accomplished (was it done with joy, or as work?). Plus we get to observe our emotional states.

Hopefully we become aware as hosts how we may be judgmental towards someone who has never gardened, or worked with tools. Maybe the wwoofer has never done manual labor...are we, as hosts, mindful of a wwoofer's needs?

Chapter 2 (Checkin) plays such an important role in being honest with our wwoofers. There have been times when we felt taken advantage of because we were so "flexible". It became a strain on forming a relationship. What develops is an US and THEM mentality.

Just like in the garden, we like to accommodate wild plants, we're "flexible" with that, however we set boundaries. Some Aspen trees are welcome, not an Aspen grove! The Caragana's pop up everywhere, so the young plants need to be pulled and reused as compost or put in pots to give away.

We encourage wwoofers to enjoy the area, so our daily schedule is flexible, the choices of daily tasks are flexible, and mealtime is somewhat flexible, all within certain boundaries. The way to building a resilient community is through consideration and compassion of "the other". The whole experience a wwoofer might have can impact them in such a way that they impact others in their sphere of influence once they leave.

As hosts what we can offer is an opportunity for a wwoofer to experience the safety, acceptance and consideration of the host, our Mother Earth, and our entire planet.

There is a lot of responsibility as a host. Especially if your intention is to host an educational farm/micro homestead

Implementation Journaling Questions

What is your intention for inviting WWOOFers to your farm?

1. Are you willing to develop a meaningful relationship with the

WWOOFers?

2. What boundaries will you establish?

3. Will you have rules? Or will you discuss boundaries if you decide to include check ins?

What might you do to establish clarity?

1. Will you have a daily task board hanging on the wall?

2. Will you designate priorities?

3. Will you let the WWOOFers choose or will you assign them tasks?

Open your heart to compassion. Be willing to listen and empower a WWOOFer to use their deductive reasoning skills with no fear of failure.

Chapter 9

"Let's Play a Game!"

How we help each other vs compete against each other

Wayne Dyer ("Father of Motivation") once said, in one of his works, that his Guru told him to "stop taking yourself so damn seriously"!

And so, we like to play. Some card games are a great exercise in becoming aware of how we fluctuate between competing against or helping others. We like to play games like Golf, or Skipbo. There's a dice game we play Greed.

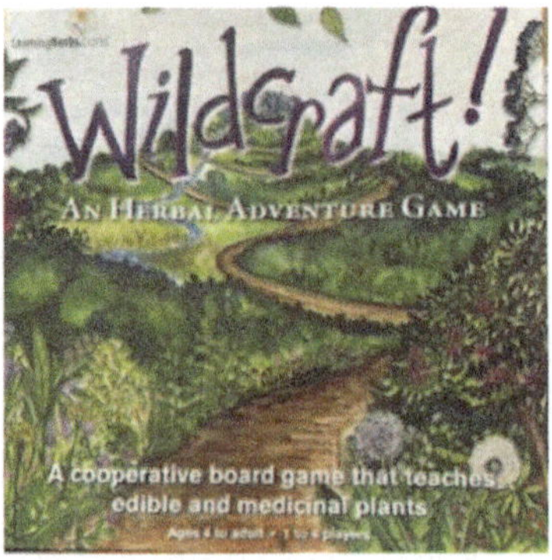

And a permaculture card game for "Living Landscapes" and garden design.

There's a "Wildcraft" board game where the participants all play to identify plants. A win-win situation!

Games are a way to become better acquainted. A personal connection might not be noticeable. However, we can get a better sense of a WWOOFers personality by the way they play the game.

Implementation Journaling Questions

1. What does body language indicate in someone?
2. What games are good team games? Aka games where all of the players are working together cooperatively?
3. If game night was something you offered to WWOOFers, how would you go about it?

Chapter 10

What will you do after WWOOFing?

Create community wherever you are

As the wwoofers leave, we realize the impact the short relationship has had on our well being. Be it a farm, garden or homestead there is always rebuilding that takes precedent over everything else. Whether it is an erosion problem, or a structural situation, even rebuilding the soil and the garden beds - everything. In our personal relationship with our selves there is rebuilding going on. And as we rebuild our inner workings, the impact can be felt by our community of friends and others. Whatever rebuilding goes on with the earth we occupy, all life is impacted.

So as the wwoofers leave us and we stand in the driveway waving good bye, we hold onto each other, tightly. Sometimes with tears in our eyes, we still feel the time spent with these special people. The personal barriers we had to break through, and rebuild in these short lived relationships remain held together by a thread of oneness. Just as the community of trees, plants and the animal kingdom, we are part of a matrix. "We wish you would stay" is a thought form or message to stay in our hearts with unconditional love for all that is.

We believe that the impact will be felt and inspire others to co-create a web to constantly connect with nature and people. When we become fully aware of how our connections and feelings of the other are intertwined we feel different. We start feeling more part of the whole. We feed one another's souls and integrate all of our experiences to offer support for each other. Age barriers cease, so do judgements, then gratitude for each other evolves. This amplifies joy and happiness for all of life.

Our WWOOFing experience has been nothing short of an educational opportunity for us to gain knowledge from a source much greater than us. To the founders of this grand idea we say thank you. To the participants that have come through our home, we say thank you. And to all who host, now and in the future, we say thank you for keeping this thread of oneness connected to the earth.

The WWOOF experience can be a very impactful experience for both the WWOOFer and the host.

Implementation Journaling Questions

1. What might a host do as follow up with WWOOFers? Occasional emails, or phone calls? Maybe a newsletter keeps a connection. How about a Sub Stack subscription?
2. Have you considered that WWOOF hosts are a part of a network that supports a transit community? A place where volunteers get a taste of a cultural movement that regenerates the earth and people towards a more cooperative society? How do you envision yourself as a host with this in mind?

Chapter 11

Epilogue

1/25/2025

As we began to write about our decade of our dedication to WWOOF as host, there was a phenomenal event happening. Six planets aligned in a straight line to the Sun. And Earth is one of them. This is supposed to be a good time to journal what it is one desires to happen during her/his life. Whatever that might be, wealth, health, happiness, you describe it, visualize it as if it has already happened and be grateful to live it.

We have this story to share and the wholeness of this experience is connection. Not just with the Earth Mother and it's entire kingdom, it includes forming relationships. I observe Marian's effort to stay connected with many of the visitors we've had as possible. Just tonight she was so excited to finally have a long phone call with Samantha, a certified mermaid and nanny. It's exciting when some visit us years later. Jake has been a regular fixture in our conversations and memories. He showed up for a couple of days and shared his life since the time we knew him as one of our first woofers. There was the artist/ woman hockey player from our home town of Chicago. She stopped in Livingston on her way to a teaching gig in Missoula and asked if she could come meet us for lunch. Her garden mural still graces the front gable of our home. Ben the fire twirler who put on a show for us one night, keeps threatening to show up, he will some day, I feel it. Maggie came back with a friend. They were here for the Bison hunt that happens in the fall.

So why are we sharing these stories? Because for us more than anything relationships mean a lot to us. They are not always easy, there is so much to work through, all of our perceptions and judgements of each other. It is worth the effort and a purpose to live for. Connecting with others, especially with a generation that sometimes is fifty years younger than us makes being a host worth the effort. The reward of giving and receiving is a grace that warms the soul.

As this special planet alignment occurs what we desire most is to be part of a Oneness movement that connects generations to the earth and pure consciousness.

> *"Enlightenment is when a wave realizes it is the ocean"*
>
> - Thich Nhat Hanh

WWOOFER Testimonials

Nicole

United States • April 2022

Marian and Jim are like a hug to your soul! They hold morning check ins over green drink to check in on you but also to honestly and vulnerably share their emotions and energy. They hold space for you to share and have your need met. This often turns into great conversation and realization about a variety of topics that lead you to grow and heal. In my short stay I healed from much, was loved deeply, and was inspired to live more truthfully and compassionately. Jim and Marian respect your boundaries completely and do not force anyone into anything but if you choose to partake in sharing deeply or meditation you will receive a gift of their wisdom and kindness. The food is lovely, nourishing, filling, and delicious. I learned a lot about health, herbs, and healthy recipes. The farm work is awesome. Their ecosystem is a beautiful example of restoring soil, growing wild, and loving your plants. I learned a lot about seed collection, dead heading and building soil. Not only that, you are surrounded by the most beautiful mountains everywhere you look! Jim and Marian have wealths of knowledge to share about any of your interests and make an effort to connect you with your passions in farming. The tiny home is cozy, comfy, and cute. I slept great! All in all I highly recommend a stay here for your farming knowledge, to receive great energy for your soul, and to get to know two lovely humans! I can't wait to come back to visit!

Finn

United States • July 2023

Jim and Marian were fantastic hosts! Very clean, organized, and knowledgeable people. Their garden has more variety that I've ever seen in my life and they both know so much about all their plants. Morning meditation was calming and a great way to start the day and evening walks were a peaceful end to the day. Highly recommend Snowbird to anyone hoping to learn more about permaculture and working a garden!

Matt

United States • August 2020

These folks are the bees knees. Jim and Marian will teach you a world of knowledge if you are prepared to learn. It's one of those things where the energy you put into this lifestyle is the energy that it will give you back. The summer I stayed gave me the opportunity to greatly improve my mental and physical health and peace, and to 'find my light' on my life path. Marian has recipes for great meals, Jim's got the dessert recipes, and both have "recipes" for physical and spiritual health. There are tons of beautiful hikes and friendly neighbors nearby, and a lot of gardening and exciting yardwork projects to do. I hope you are able to fully appreciate the scenery and the opportunities offered by these wonderful and wise people!

Unaa and Sarah

Belgium • Summer 2014

How can we imagine a better stay that we had with James and Marian !?? We learned about ourselves and the way to live in community.

Veronica

United States • September 2024

My time with Marian and Jim far exceeded my expectations for a WWOOF stay. They are incredibly kind and taught me so much about mindfulness, intentions, and gratitude, in addition to sharing their gardening and cooking knowledge. I appreciated their invitation to embrace the present moment along with whatever emotions it holds. If you need another reason to visit, the surrounding area is stunning. Waking up and seeing the mountains every morning made it easy to remember the blessings that life offers. I feel extremely grateful for the time I spent here and for the new "family" that I gained.

Takayuki

Japan • November 2016

We had a such a wonderful time in here. Jim and Marian were very kind to us and always make us peace of mind. Just like who we wanna be. Marian made us healthy and tasty green smoothie every morningthen we check-in which is conversation and meditation each other at the beautiful back porch with the mountain and the sun. Of course we learned a lot of permaculture things as a fun work. When I bring back that memories, I feeling what a peaceful place. Thank you for let us stayed there and learning and delicious food and playing the games. We love you guys so much.

Well, we also want to say... Hear O Universe, I am grateful! TAKA and KONOMI

Desiree

United States • October 2015

Montana would not have been inspiring without Jim and Marian. Ryan and I stayed in their loving home and helped maintain their garden. Each morning we would wake up with Marian and have a homemade green drink (directly from their garden) along with check-in. We miss the uplifting check-in's each morning with you Marian and Jim! Ryan and I find ourselves communicating more with each other about our day and starting each day with our own green drink :) After check-in, we would go outside in the garden and water, weed, slug patrol and learn through Marian and Jim about Permaculture. When we were done we would adventure around Paradise Valley and do mini field trips to the locals in the area. We learned about Monarch Butterflies and how to raise chickens! Everyone in Paradise Valley was very friendly, humble and amazing. When we left we brought many lessons with us we will never forget and we have Jim, Marian, and all of the empowering people we met to thank for. Love you guys!

Jacob

United States • March 2015

Jim and Marian are two people that are simply a gift to humanity. They have a love and commitment to being beautiful to the earth and others that is unparalleled. My time there was highly transformative for my own self and I learned so much about the Permaculture philosophy, which now, because of my time spent with Jim and Marian, is my life's passion. If you have a chance to WWOOF anywhere, WWOOF here. Their home location is jaw-dropping beautiful and living in community with these two is equally beautiful.

Contributors and their Stories

Veronica

The last farm guest of 2024 — What a blessing to know her

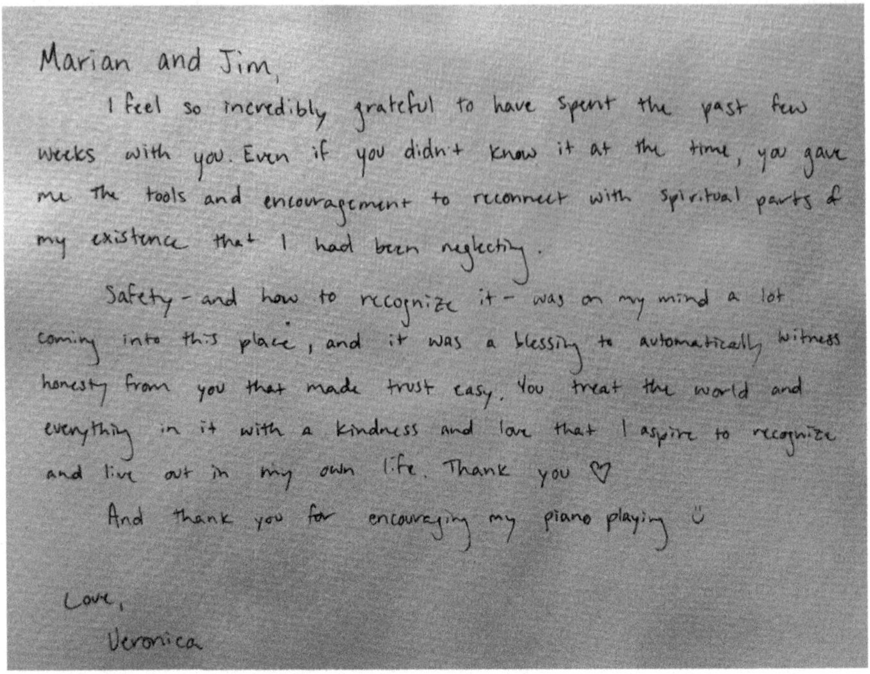
Marian and Jim,

I feel so incredibly grateful to have spent the past few weeks with you. Even if you didn't know it at the time, you gave me the tools and encouragement to reconnect with spiritual parts of my existence that I had been neglecting.

Safety - and how to recognize it - was on my mind a lot coming into this place, and it was a blessing to automatically witness honesty from you that made trust easy. You treat the world and everything in it with a kindness and love that I aspire to recognize and live out in my own life. Thank you ♡

And thank you for encouraging my piano playing ☺

Love,
Veronica

Jake Desatoff

"My experiences wwoofing were deep and beautiful. The connections I made with fellow farmers were powerful and transformative. Working on people's properties in exchange for help created unique bonds and often lifelong friendships.

I initially thought farming was all about collaboration, but I soon realized it requires a lot of commitment and resources. It's tough work, and it needs a heartfelt connection to the earth, animals, and the community. Being out in nature while tending to the land gave me so much in return.

When I signed up for wwoofing, I wanted to learn about sustainable practices like permaculture and how to produce food sustainably. My goal was to understand these concepts on a homestead scale, not for large-scale agriculture. I was pleasantly surprised by how you engaged with your space in a regenerative, thoughtful way.

Wwoofing changed my life and motivated my wife and me to travel around the Western U.S. in a camper, exploring communities focused on agriculture. We aimed to find land to start our own farm.

My time with Jim and Marian transformed my spiritual journey. They helped me connect with a community focused on intentional living and self-awareness, which led to healing from past traumas. Our bond has lasted a lifetime, and I'm grateful for the deep connections formed through wwoofing.

I also improved my skills, especially in soil quality and fertility. I became fascinated with composting and soil biology, realizing it's the foundation for successful crops. Learning about site design and how to efficiently manage a homestead was another valuable takeaway.

If you're thinking about wwoofing, go for it! Make room in your life for this adventure, whether it's for a week or longer. It will positively impact you, offering challenges, lasting relationships, and possibly a new life path. So, if you feel that itch, listen to it and embrace the experience. You won't regret it!"

Veronica

Her recollections from her year-long woofing experience.

Many of my favorite books and movies are ones in which the main character encounters a seemingly rudimentary creature whose simplicity proves to hold immense wisdom. Over the course of these stories, the creature acts as a mirror to the main character in a way that reminds them of their own humanity and of what really matters in their one precious life. Little did I know when I started WWOOFing that I was about to spend the next 14 months becoming acquainted with countless people, animals, and plants who would hold up a mirror to my own existence and invite me to seek out the meaning in all of it together.

After graduating college, I chose to travel and participate in WWOOF because I no longer felt sure that the field I had just spent 4 years studying in was where I was supposed to be. Amidst a lot of uncertainty about my future, one thing that felt important to me was to learn how to grow food and give back to the land, so that's what I did. At each of the 15 places I stayed, I met people who had similar uncertainties about their own lives. Regardless of whether we were a few years or a few decades apart in age, we were all in periods of transition where we were taking steps back and reevaluating what was important to us. After a WWOOF host in Louisiana asked me what I was doing all of this for, I realized that I was on somewhat of an unintentional quest; I was searching for wisdom in the ways people were choosing to live. At some point, I had stopped asking myself what I wanted to do with my life and was now asking myself how I wanted to live.

During my travels, in addition to staying with Jim and Marian in Montana, I lived with a collective of artists, musicians, and circus performers in Washington, a group of wild-spirited builders in the desert of west Texas, a spunky chosen family situated in the Redwood forest, and a compassionate and endlessly inquisitive couple in the Black Hills of South Dakota. I was able to witness people living in ways I had never considered possible. Time and time again, I was shown examples of how extraordinary, finite, and cyclical life is. One beauty of farming, it seems, is that as we cultivate the lives of plants and animals, we are simultaneously learning to savor our own. These experiences taught me the importance of letting myself be led by awe, curiosity, and delight. The future will always hold uncertainty, but if I continue to look to the land and those tending to it for guidance, I might be able to find meaning wherever I am. Maybe one day I might even be the creature reminding someone else of their exquisite and precious existence.

Anna

Her letter detailing how she almost didn't come, but then she listened to her intuition.

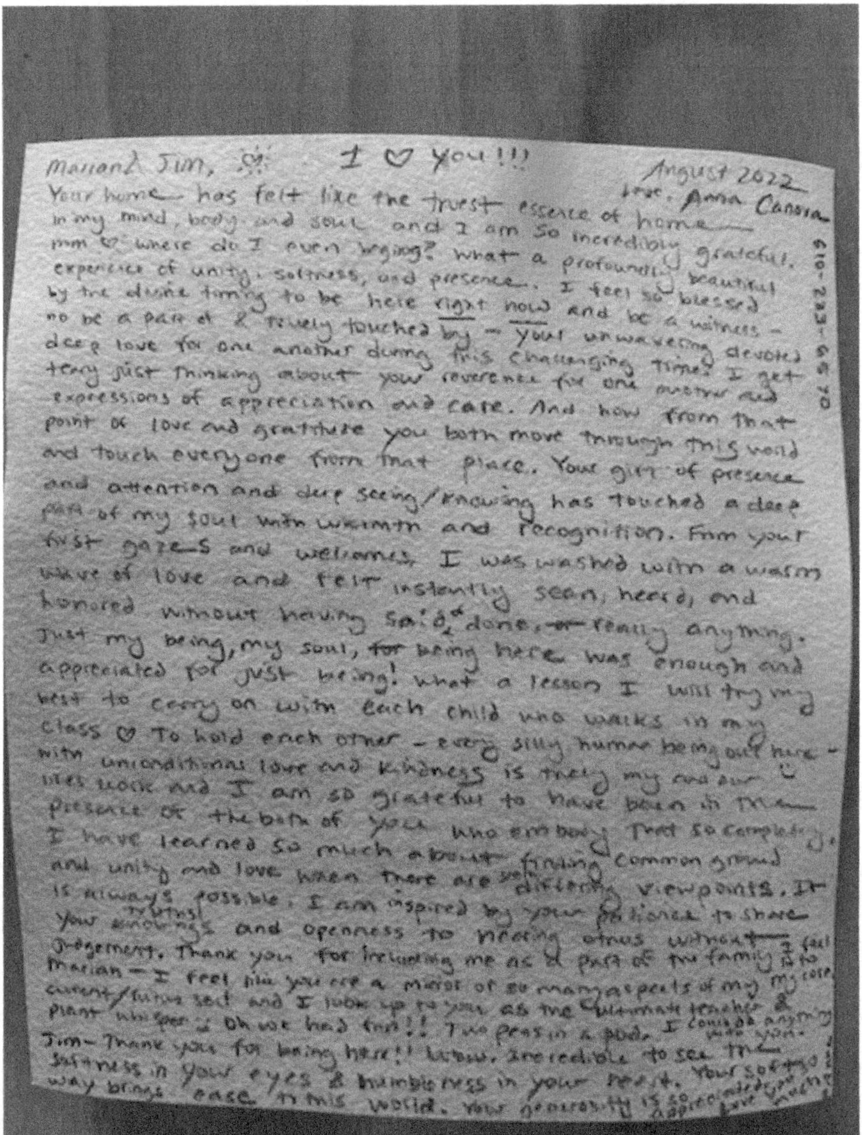

Marian & Jim, ☼ I ♡ you!!! August 2022
love, Anna Canora
Your home has felt like the truest essence of home
in my mind, body and soul and I am so incredibly grateful.
mm ♡ where do I even begin? What a profoundly beautiful
experience of unity, softness, and presence. I feel so blessed
by the divine timing to be here right now and be a witness –
to be a part of & truly touched by – your unwavering devoted
deep love for one another during this challenging time. I get
teary just thinking about your reverence for one another and
expressions of appreciation and care. And how from that
point of love and gratitude you both move through this world
and touch everyone from that place. Your gift of presence
and attention and deep seeing/knowing has touched a deep
part of my soul with warmth and recognition. From your
first gazes and welcomes, I was washed with a warm
wave of love and felt instantly seen, heard, and
honored without having said or done, ~~or~~ really anything.
Just my being, my soul, ~~for~~ being here was enough and
appreciated for just being! What a lesson I will try my
best to carry on with each child who walks in my
class ♡ To hold each other – every silly human being out here –
with unconditional love and kindness is truly my and our
lifes work and I am so grateful to have been in the
presence of the both of you who embody that so completely.
I have learned so much about finding common ground
and unity and love when there are seemingly differing viewpoints. It
is always possible. I am inspired by your patience to share
your ~~knowings~~ truths! and openness to hearing others without
judgement. Thank you for including me as a part of the family. I feel
Marian – I feel like you are a mirror of so many aspects of my my core
current/future self and I look up to you as the ultimate teacher &
plant whisperer. Oh we had fun!! Two peas in a pod. I could do anything with you.
Jim – Thank you for being here!! Wow. Incredible to see the
softness in your eyes & humbleness in your heart. Your soft gentle
way brings ease to this world. Your generosity is so appreciated. Love you so much!
610-283-6590

A.M.

A fellow host of wwoofers recounts a special relationship with one of them.

> "My woofer was an Israeli Russian emigrant to Canada who visited my place in the deep forest where I was living in a 9 by 14 shack in the middle of winter and building a small house. Later on, she called me also in the middle of winter when the shack became available. She asked if she could come and stay for a couple of months. This is freezing, middle of winter, Wilderness Canada. I said yes. She was a wonderful worker. A wonderful person and she worked for me while I stayed in the other place that was being under construction. We became friends she moved into the area later on. LanarkCounty, Ontario. She loved the wilds, partly because of her Russian childhood. We're friends to this day. It was a very, very positive influence in both of our lives. We have about a 50 year age difference and we became friends despite that. I'd be happy to tell much more if you'd like James and Marian. If you'd like to talk to me about it, call me. Thank you."

Appendix

"WWOOF® links visitors with organic farmers, promotes an educational and cultural exchange and builds a global community conscious of ecological farming practices.

Visitors (or "WWOOFers") participate in the daily life of hosts while receiving educational opportunities, meals and accommodation during their visit, with no money exchanged between hosts and WWOOFers. WWOOFers learn to grow their own food, help on the farm, and experience a new culture. Hosts open their homes to receive visitors from all walks of life who want to connect with organic food, regenerative agriculture and give a hand in the everyday tasks. Embark on an adventure and discover unique places, where growing food from hands-on farming is part of an inspiring lifestyle. Live in-tune with the elements of nature, where there is good laughter around the table and real food on the plate. Team up with WWOOF to promote organic agriculture and support farmers who are nourishing their community." wwoof.net

WWOOF CHARTER

ARTICLE 1 - WHAT IS WWOOF

WWOOF is a worldwide movement linking visitors with organic farmers and growers, promoting an educational exchange and building a global community conscious of ecological farming practices.

ARTICLE 2 - WWOOF AGREEMENT

Hosts welcome WWOOFers and offer them clean, dry and safe accommodation, as well as sufficient food for three meals per day.

WWOOFers express a sincere wish to discover the hosts' lifestyle, to join them in their daily activities and respect the location and the people there.

ARTICLE 3 - PARTICIPATION

WWOOFers help out with farm activities in a free and voluntary manner for the agreed upon time per day. The exact daily schedule may vary from host to host and is agreed upon prior to confirming a visit. WWOOFers have no obligation to be profitable and are in no way subordinated to the hosts. They receive no pay and must in no case replace an employee.

ARTICLE 4 - COMMUNICATION

WWOOFers and hosts prepare for visits by discussing activities, schedules, and expectations beforehand. Communication is maintained until arrival. Members commit to answer all messages, even to decline a request. Cancellations must be made with as much advance notice as possible.

ARTICLE 5 - EDUCATION

Hosts express a sincere desire to share their expertise in organic farming, agroecology and their ecological lifestyle. They have an informal educational approach and demonstrate by example how to live in harmony with their environment.

ARTICLE 6 - AGROECOLOGY

Hosts cultivate the land using organic and ecologically sound techniques. All members support and encourage agroecology and are committed to a sustainable way of life.

ARTICLE 7 - SAFETY

WWOOF organizations do not tolerate any kind of legally recognized discrimination. Nor do WWOOF organizations tolerate any form of sexism, harassment, or sexual violence. They are committed to providing an inclusive and welcoming environment for all members, and require all hosts and WWOOFers to abide by this policy.

ARTICLE 8 - COMMUNITY

All WWOOF members are encouraged to leave an honest review about their experience to help strengthen the WWOOF community. Both WWOOFers and hosts are invited to report if member profiles are inaccurate or if members' actions violate the WWOOF Charter or terms and conditions.

ARTICLE 9 - INSURANCE

Everyone must be properly insured for the risks inherent in their activities. Hosts are responsible for assessing risk involved in tasks, providing adequate protective equipment and training where necessary, and not involving the WWOOFer at all if the risk is high. WWOOFers are also responsible for assessing risk and keeping themselves safe.

ARTICLE 10 - PRIVACY

WWOOF organizations undertake not to use their members' data for commercial purposes. Members agree not to reproduce or disclose the host list or to use it for anything other than WWOOFing.

WWOOF FoWO Federation of WWOOF Organisations

Wwoofers - We cherish our decade of connection, thank you all. There are many photos that are missing, you are all imprinted in our hearts.

Above left part of the front yard about 7 years ago.

Above right, the yurt in full bloom.

What is the posture?

It is done in seated position. You may either be seated on a chair or on your meditation cushion on the floor. Your palms resting on your thighs, you keep count with your fingers at every stage of the meditation. Your thumbs gently touching the tips of each of your other fingers.

The 6 steps of Soul Sync:

The first stage of Soul Sync - you practice 8 conscious breaths. As you move from one breath to the other, you keep count with your fingers. It is okay if your attention wanders, please bring it back and continue unperturbed.By the time you finish the first stage in Soul Sync, your para -sympathetic nervous system is fully active. Therefore your blood pressure begins comes to balance. And your digestive system begins to be impacted positively.

The second stage - is to inhale deeply, and while exhaling you make the humming sound of a bee at a base pitch. Please listen to the sound with total attention. When you do this, spiritually your ajnya chakra is getting activated.It is releasing nitric oxide from your sinuses, which will cause a deep sense of relaxation within your body and also positively impacts your heart muscles. Nitric oxide when released slows down your pulse. It leads you to a much deeper state of relaxation.

At the third stage - You observe the pause between inhale and exhale. When we inhale and exhale, there occurs a natural pause after each inhalation, before exhalation begins.Observe that pause. You do this for 8 cycles of breathing. As you do it, you begin to experience a slowing down of your thoughts. It is a beautiful spiritual experience.

In the fourth stage – You inhale and exhaling as you inwardly chant the mantra "Aham".Aham means, "I am Limitless Consciousness". You repeat the mantra 8 cycles. At this stage, you move beyond your limited self. You realize that you are infinite conscious-ness. You are taking your meditation beyond calm to expansion.

In the fifth stage - You imagine your body expands into light, into a field of consciousness ... the floor, the table, and the people around you ... everything expanding into one unitary field of consciousness. You can either feel it or think it. In this field of conscious-ness, everything is connected. Everything is one. It is one single unitary field of conscious-ness. There is no separation, no division. In this field, thought and matter are one. Desire and reality are one.

In the sixth stage - you feel or see or imagine one of your heart felt desire. Feel it, as if it is happening in the now. Feel that intention's manifestation through all your senses. See it vividly, hear it, smell it, touch it and taste it.Let the future be here in the present moment.And after you have done your soul sync go forth and act in accordance with this intention.

Influential Books

The Permaculture Promise

A very detailed book the impact of permaculture has for regenerating the earth - rural and urban.

The Backyard Homestead

Great book that covers a lot and gives one the structure to plan for building a homestead. Includes ways to process the food you grow and whatever animals you raise for food self-sufficiency.

Together Resilient

I really believe that building small sustainable communities is an alternative to the "helter skelter" of the "other" illusionary world of competition and conflict. So this book hit home, and for us the Wwoof program is a way we can contribute to inspire.

The New Seed Starters Handbook

This is a book we have had for a longtime. It's everything you need to know for raising healthy seedlings successfully. More than anything it preserves growing organically and guarantees use of non GMO seeds.

Foraging The Mountain West

This was one of the books we received in the trade for the wood burning water heater that we mentioned in one of the chapters. This book reminds us of being members of Foraging Friends when we still lived in Chicago.

Designing with Nature

Jim took this course with Cal Earth on how to build homes with sand bags. Labor intensive, low cost and sturdy. Plus the design has so many options to be free flowing. There was a small village constructed with drainages and garden spaces.

Natural Materials

Another Cal Earth video with Nader Khaliili. His designs were approved by the county building department. They withstood the forces of an assimilated earth quake and more. When I took the course you could purchase stamped architecture.

Inhabit

I'd have to say one of the best movies I've seen that captures the permaculture concept. That is taking a piece of land that is of little value, and turning it into a precious piece of earth.

Keeping Bees with a Smile

This book is great resource for beginners who want to keep bees and for beekeepers who want to consider other natural alternatives. Wwoofers have made "bee condos", insect motels, and every year when the Holly Hocks dry the hollow stems can be used to replace the rooms in the hotel! The Magpies use them for their nest. We also have many plants that attract bees and a wasp nest in a hoop house or under an eve don't bother us since they assist the garden as pollinators. We will sit in our sunroom and watch the various bees that visit our garden. It's quite a mesmerizing time spent with bees, birds, butterflies and other insects. It surely puts a smile on our faces.

How "Native Americans" Use Wild Plants for Food, Medicine & Crafts

The title says it all. The book is about the indigenous Chippewa of the great lakes region in Wisconsin and Minnesota. It's a detailed book of ethnology with old photos and plenty of charts and references.

Secrets of the Soil

"Tompkins and Bird, coauthors of The Secret Life of Plants , cite the work of scientists and philosophers such as Steiner to support their extravagant claims; they report on the special qualities of glacial water and efforts to produce it in the laboratory, and tell of a plant breeder who believes that birdsong, music and a natural foliar fertilizer can stimulate plant growth ("sonic bloom"). Such exotic journeys into the fringes of agriculture, best appreciated by true believers, make ordinary organic farming and integrated pest management, though perhaps more effective, seem positively mundane."

Gardening for a Lifetime

Ha Ha, thanks to wwoof we are able to manage a garden that gets larger every year.Thanks to this book we have a sense of aging in grace as gardeners.

Build with Adobe

Another labor intensive natural way to build using the dirt from the earth. Jim had lived in an original Adobe home in Rancho de Taos across a large vacant lot from the Mission. It was warm and cool just like that style of home is supposed to be. We have this book since Jim had planned to build this way in New Mexico. It never happened but you never know with him...maybe an outdoor oven? Or suana? Gosh, I wonder what he's going to do with the roll of sand bags he purchased from Cal- Earth? I hope he reads the Gardening for a Lifetime book and realizes he's getting older. Maybe, a Wwoofer will take up the project. Someone actually started too. But now we have a garden plot in the area. Much easier and more practical at this time in our lives. HA HA

Wolves and Humans

by Elke Duerr an animal activist who visited us often. Https://vimeo.com/weboflife/Videos

Round America with a Duck

A 1960s-child (now a married mother-of-two) about to turn 60 travels 10,000 miles/16,100 kilometers via bike, buses, trains and WWOOFing, working more than 500 hours on organic farms in exchange for free food and accommodations. Join her — and her silly little duck — as she meets America from coast to coast. You'll laugh. You'll cry. You may even get inspired to try something new on your own life journey!

Comparing Cultural Worldviews

The following chart is from the book Together Resilient by Ma'ikwe Ludwig. The book's premise is based on "building community in the age of climate disruption". We inserted this here, not to highlight climate change, rather to note the differences in the way we approach intention. We like to operate in the Sustainable (we like Viable) Cooperative Culture column. The "collaboration with allies" includes people with different views, nature as in plant combining, no such thing as weeds, and insects as in wasp and bees. A viable culture that grows despite external conditions.

Comparing Cultural Worldviews
and the behaviors that come from them

Extreme Competitive Culture	Sustainable Cooperative Culture	Extreme Cooperative Culture
Compete with others	Cooperate, including collaboration with allies	Cooperate within; collaborate only with others your group fully agrees with
Seek advantage and winning	Seek understanding and effective action	Seek attention as connection
Have skill? Use to dominate	Have skill? Teach with discernment	Have skill? Give away indiscriminately
Loudest voices win	Collaboration/consensus with discernment	Consensus with no discernment

Midwest Permaculture Online Design Course

Midwest Permaculture - A way to learn online.

https://www.midwestpermaculture.com

We can offer this online design course to our Wwoofers and up to six Wwoofers can sign up at half price in one calendar year.

Sri Preethaji – "Awaken to the Living Universe"

Youtube Link – "Awaken to the Living Universe"

"Let us understand that this is a living Universe. It is a conscious Universe. And there is communication happening all the time within each life form. Communication between one life form to another and various life forms to the whole."

"If the entire Universe is communicating, how are you communicating to this magical field of consciousness? You are not communicating through your words alone. You do not communicate through your actions and your behaviors alone. You communicate to the Universe through your state of consciousness."

Visit us at WWOOF.org and search for Snowbird Agricultural Trust.

Acknowledgements

There are several acknowledgments to give to people. Of course the publisher and folks (Matt and Amanda) at "Punchy Books". Their inspiration to keep it short and sweet and to the point was encouraging. Who needs words when you have pictures? And to the WWOOF organization, their world wide effort to connect people to the soil for the regeneration of Mother Earth! Kudos to our accountability group, Janice, Sandra and Teda, keeping us on track towards completion. We meet every Monday at 3pm to get encouragement and become organized. Keeping us on track. We give a huge hug and thanks to all of the Wwoofers that graced this little piece of property to help make it heaven on earth. We live in Paradise Valley, and when we look out the windows of the sun room and prepare for check in (with green drink in hand) who we see are the wwoofers that already contributed and those that are coming as we start our new decade of blessed connections.

Some other Author Books (also unedited monthly PDF file to read) Go to https://inspiredbookwriters.com/f/woof

The tide moved in over night, as it always does. Jesse strolls the beach as the tide recedes, as he always does. And the tide washes surprises onto the sea shore, as it always does.
What The Seashore Left Behind
By James Kozlik

When They Leave
by James Kozlik

A JOURNEY OF LOVE
CIRCUMSTANCE AND CONSEQUENCES
James Kozlik

The Essence of Existence
A Journey of Love

TAPAS
A Journey of Love

WWOOF Host and friends

Marian pictured below

Made in the USA
Monee, IL
03 November 2026

67377510R00056